Building a Race Car
Picture-By-Picture

by
Steve Smith

Editor Steve Smith
Associate Editor Georgiann Smith

Published by

Steve Smith Autosports

P.O. Box 11631, Santa Ana, CA 92711

A Special Dedication

This book is dedicated to race car builders and crafters everywhere, in hopes that the ideas presented here will help make their task and product better. In particular, this book is dedicated to a few men I have known whose craftsmanship and race car building acumen are first class, and have contributed much to the racing world in their own way: Ivan Baldwin, Frank Deiny and Tex Powell.

Steve Smith

The two large cylinders on the firewall are reservoirs for brake fluid for the disc brakes. This car uses the small, compact Girling master cylinders, which are found on most European passenger cars in various piston sizes. Notice the triangulated brace for the hood hinge bracket, and the racer's tape which lines the bottom of the windshield. The tape helps prevent cracks from developing in the windshield because its edge is highly stressed.

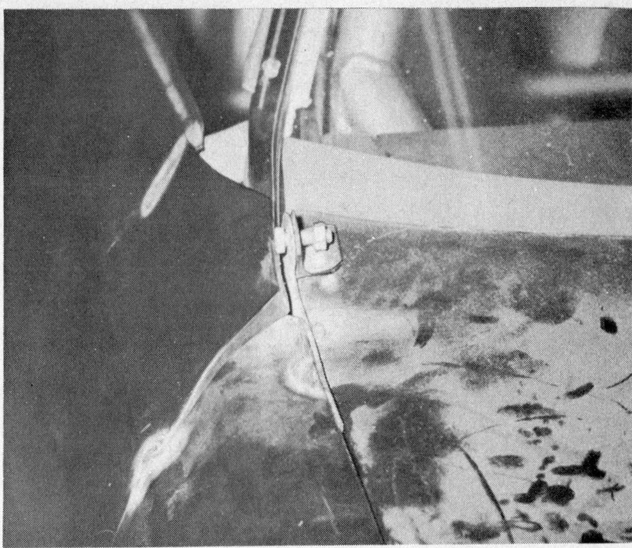

This hood hinge is improperly installed because it doesn't attach to enough material. Vibration will break its mounting loose.

See the neat throttle linkage? Just drill a nice little hole, bolt it on and you're in business. What's this piece from? A 1961-63 Buick Special. Notice the starter button on the firewall—great for adjusting valves. We must point out the balance bar setup in which the master cylinders are set. This is a mighty heavy and long piece to not be supported at the front. Bet the brake pedal gets spongy!

Putting a square tube subframe on a Chevelle Chassis isn't all that hard. And, the union of the two pieces at the Chevelle frame's perimeter kick-out actually make it stronger than a full square-tube frame. See the diagram on page 3 and instructions below for the method of mating. (Drawing on page 3 courtesy of Speedway Engineering.)

To install frame snout on most cars, cut off stock front section leaving enough so frame snout can be fit between stock frame. Be sure to plate and reinforce spliced area.

Figure 1. TOP VIEW

To determine alignment, draw a straight line from outside edge of the frame. (Line should extend beyond new front frame section.) Measure dimension A-1 (frame line to motor mount bolt hole).

Frame should be adjusted until A-1 and A-2 dimensions are the same. To recheck alignment, measure dimension B-1 (outside frame member to frame line). Dimension B-2 should be the same.

To determine dimension C, measure from the center of lower A-frame mount to center of rear end to determine wheel base.

Figure 2. SIDE VIEW

Make sure frame is absolutely flat and level on the floor. When the frame is centered (see Figure 1) measure top of frame member to floor (6¼"), then measure top of spring pocket to floor (12¾"). These two dimensions will determine A-frame angles, roof height, air cleaner, oil pan and ground clearance. Frame should be tack welded and all clearances checked as hood clearances and others vary due to different body styles and motor sizes.

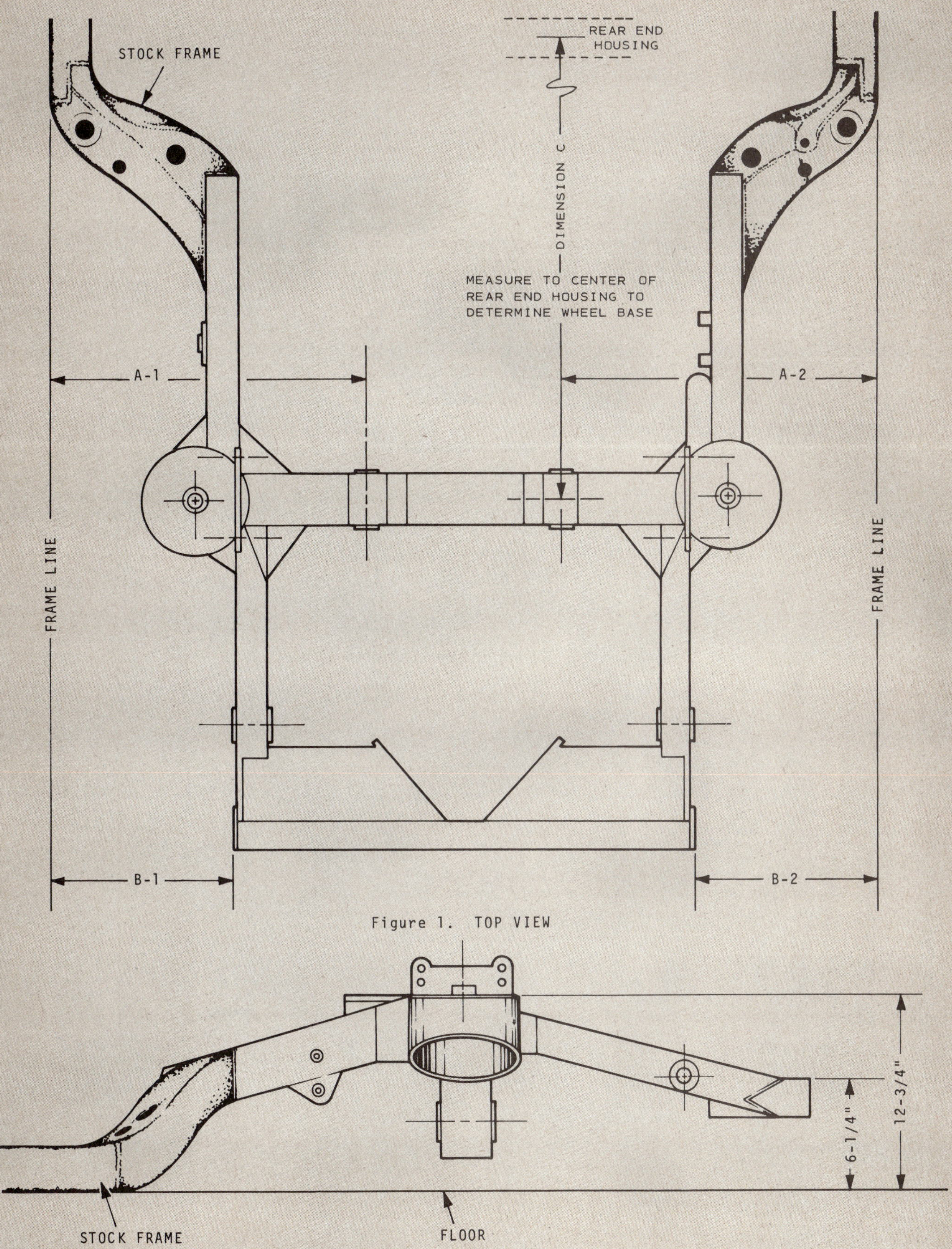

Figure 1. TOP VIEW

Figure 2. SIDE VIEW

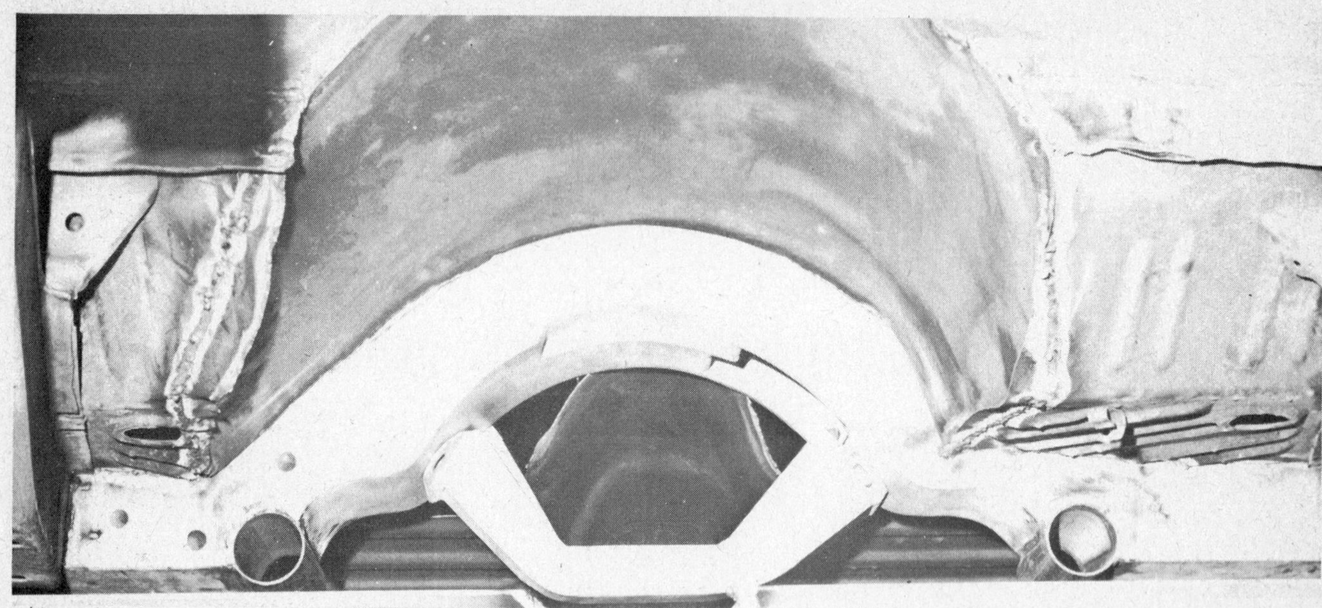

No matter what model or year of Chrysler product you use for building a race car, you should always use this crossmember which anchors the rear of the torsion bars. Nothing else is strong enough in the unibody vehicle. This is Petty Enterprises part number PE-71059.

Another view of the bullet proof Chrysler torsion bar crossmember. It weighs in the neighborhood of 125 pounds, but the weight is placed low in the vehicle and crossmember really does its job when it comes to torsional rigidity.

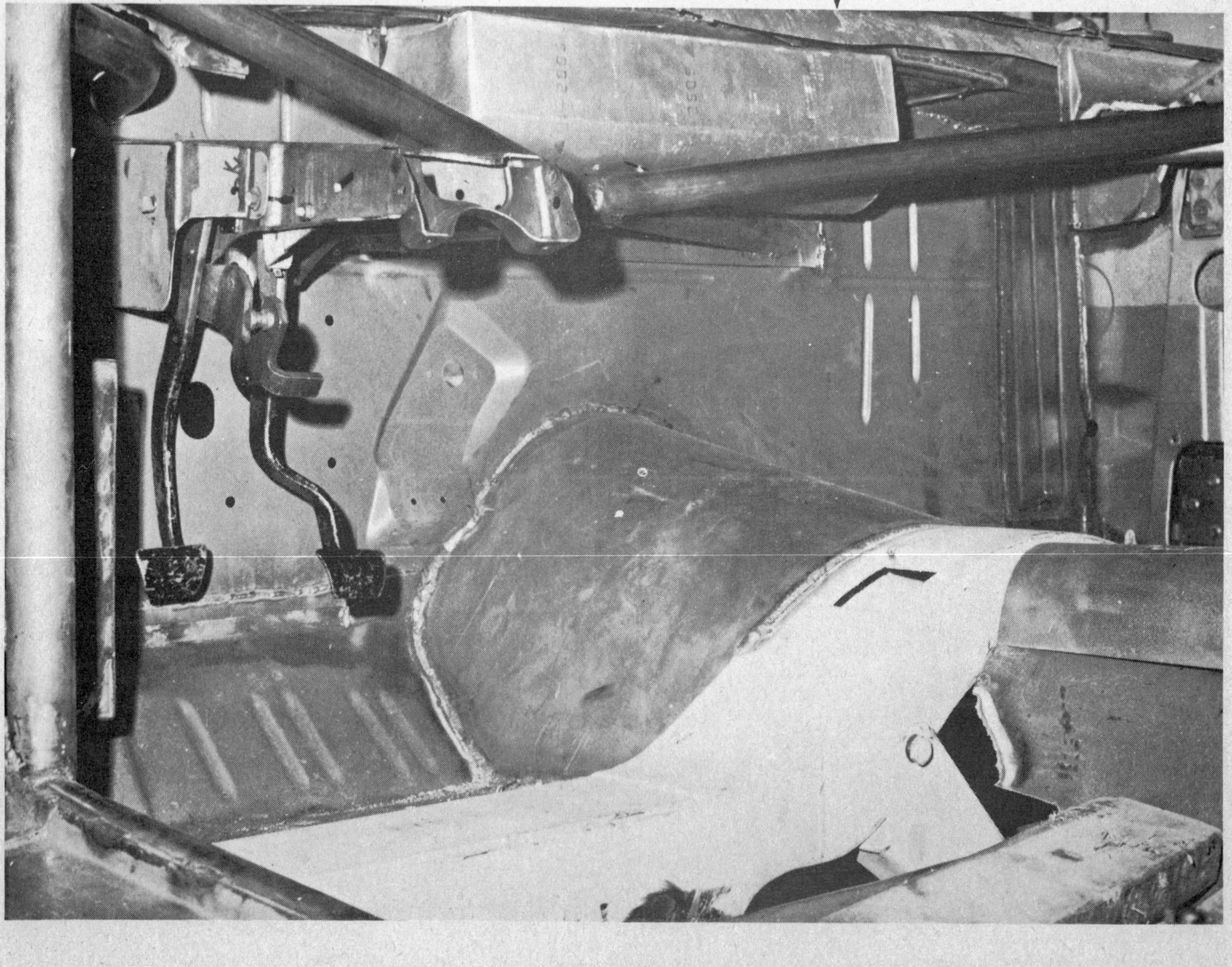

If you need to space your tie rod ends lower to correct for bump steer, consider this type of spacer. Unfortunately this hardware is custom made, but you can have it made for you too.

To set a unibody over a square tube chassis, trim the door sills until half the frame rail is covered by the sheetmetal. Weld the sheetmetal to the frame rails, then butt the floorpan up to it and weld in place. The chalk-mark circle indicates where runners from the door bars would weld to the frame.

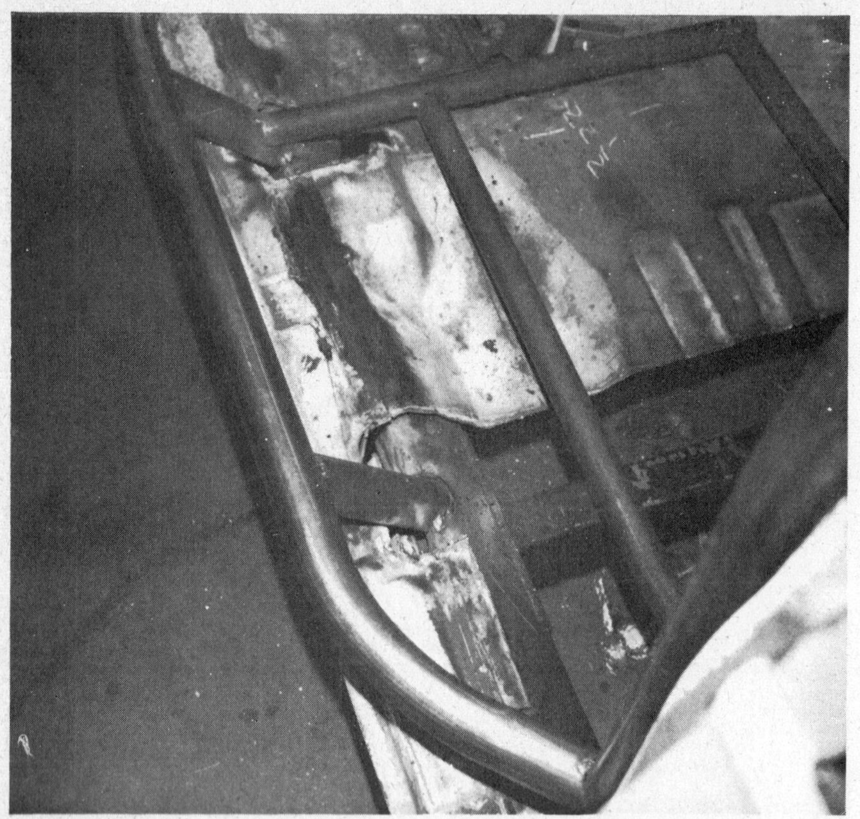

The door sills on this Chevelle body have been cut and channeled to lower it over the frame rails. Any depressed floor wells in a floorpan which has been lowered must be replaced by flat sheetmetal (if allowed by rules) or a flat floorpan such as '67 Ford Galaxy or '68 Camaro.

Below, a unibody Camaro being prepared for SCCA racing. No frame rails can be added, so the roll cage is constructed to serve as the main structural member. Notice how the front leaf spring hangers are anchored to the cage. The spring hangers are adjustable for height to keep the springs absolutely parallel to each other if running height is changed.

Above, except for a very long-armed driver, this would be a very inconvenient spot to grab the ignition in an emergency. It is our strong belief that all race cars in the U.S. should have the ignition kill switch in a standardized place so its position is second nature to all drivers.

The arrow points to the sliding adjustable bracket on the anti-roll bar which adjusts bar arm length.

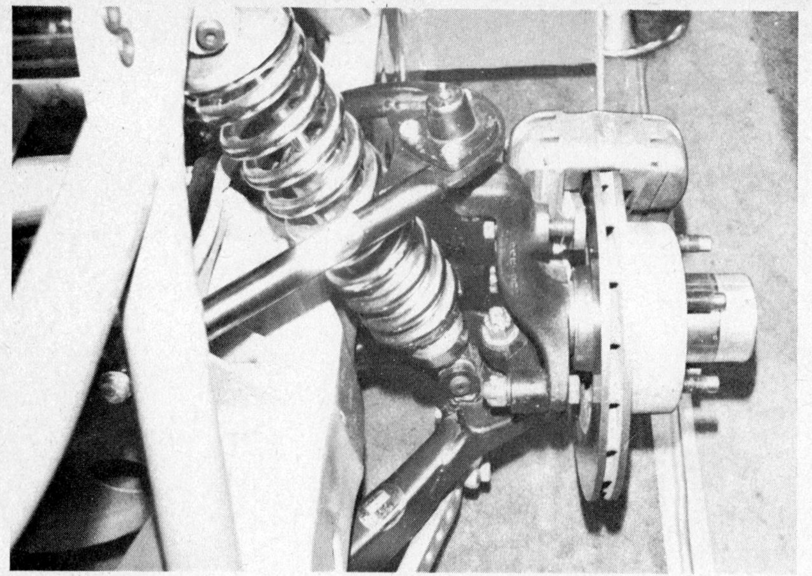

The interior section of the lower A-frame has been cut away to save weight—most importantly, unsprung weight. The bottom of the Carrera coil-over suspension unit is mounted one-inch on-center from the lower ball joint, an important dimension which allows shock and spring to control smaller wheel movements.

Looking forward into the nose of a '71 Mercury, notice how everything is completely sealed to direct the air flow into the radiator only. Even the smallest cracks are sealed with silicone sealant. This isn't just a superspeedway trick. Directing the air through the radiator helps even on the shortest track. Also notice how the projected nose of the Mercury is supported against minor bashes.

Blowing steam out the rear end? Must be a mid or rear engined car, huh? No. A piece of flexible plastic tubing runs from the radiator overflow tank along the frame rails to an exit by the rear bumper. If the radiator blows water out, it doesn't blow out right onto the front tires.

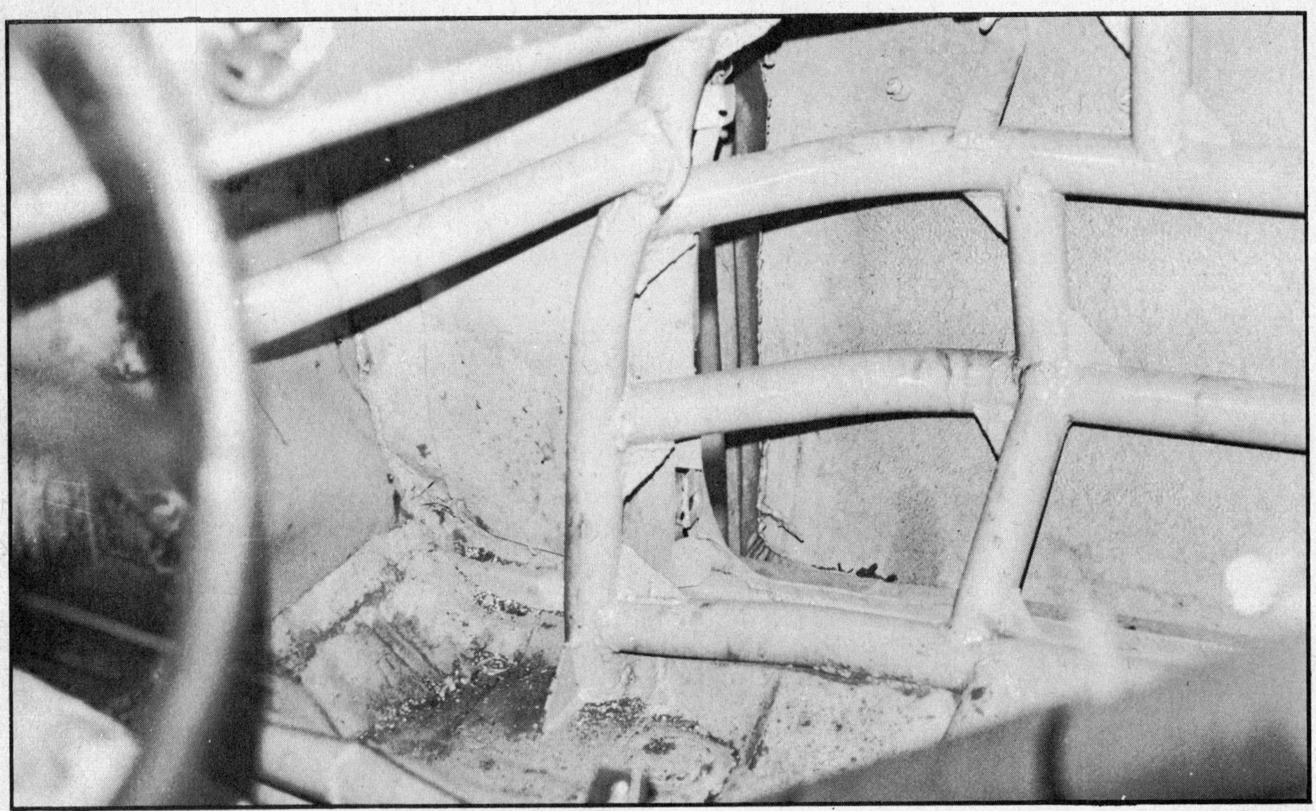

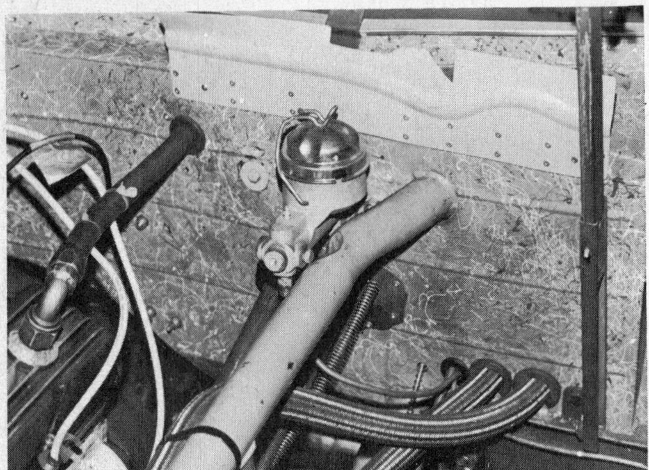

Above, this car made just a light contact with anohter car at its right front corner. But, the damage got all the way into the door bars. Why? The bar running forward from the door bars was not supported by another bar. If it had the support of a bar in back of it, there would most surely have been no damage to the upright or door bars.

At left, this illustrates a situation which is very dangerous. Undoubtedly, the placement of these two components was not planned this way, but time to remedy the situation was not available. Both the structural stiffness and the driver safety is compromised here with the notch in the tubing. This tube in question is the front hoop which supports input loads from the left front suspension, as well as protects the driver from collapse of the front end in case of a collision. The notch in the tube, combined with the radius in it, gives the structure an automatic failure point. Don't compromise the structural stiffness of your chassis for the convenient location of a component.

A good way to fine tune your handling with one anti-roll bar size is to have two alternate mounting positions to tailor arm length.

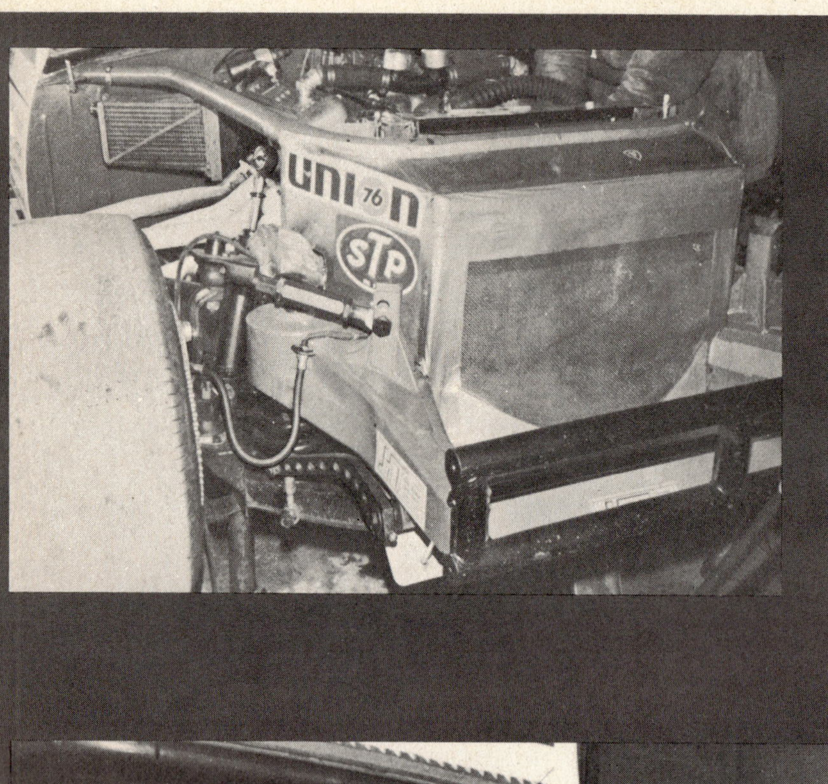

The anti-roll bar arm is attached to the lower A-frame too far from the ball joint to really get much spring rate from the bar. The upper A-arm straddle mount attachment bracket is very rigid.

Below, an interesting way to mount an anti-roll bar. Definitely rigid. Works great as long as you plan your fan location. Notice where the bar mounts at the lower A-frame. There are a series of four holes in the lower mount to adjust the length of the bar arm to fine tune the bar's spring rate.

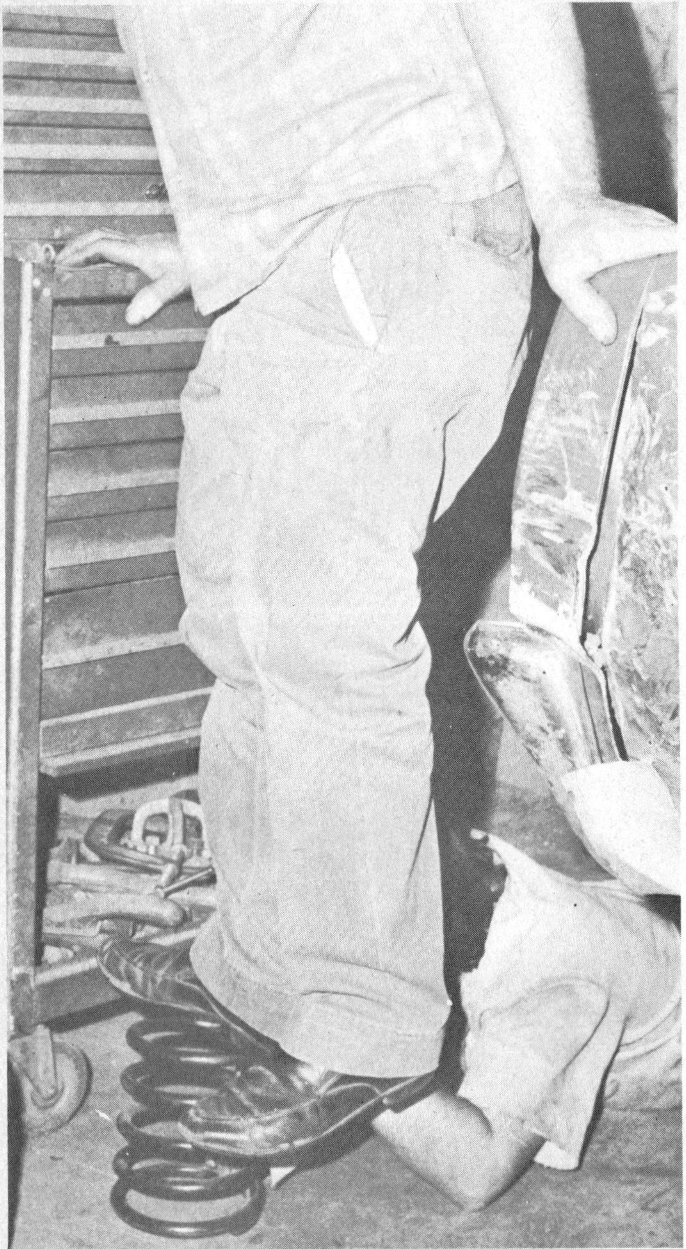

A semi-effective method of rating a spring.

A good mount for an anti-roll bar. Notice how closely it is mounted to the ball joint, and how it is gusseted.

These straddle-mount upper A-frame mounting brackets are the strongest type of mount. But they are more difficult to build, and to adjust front end geometry settings. We wouldn't suggest using Heim joint (or spherical rod end) mounts on A-frames of vehicles weighing any more than 2800 pounds.

How do you form lead ballast? How about melting the lead and pouring it in an old valve cover as a mold? Drill two holes through it so it can be bolted to brackets. To gain left side and rearward weight bias, this is a perfect place for the ballast.

This is a roll cage typical of those which so many people claim are "absolutely rigid." The use of a lot of tubing doesn't necessarily make it so. This cage is very poor, and leaves the chassis flexible. And, there's probably 75 feet of unneeded tubing in it.

These two photos show an installation of a rocker arm type of upper A-arm suspension. While it can offer some advantages in unsprung weight savings, it greatly adds to sprung weight if done properly. The upper rocker arms shown here are much more massive than they need be. The major criteria for an upper arm of this type is that it not deflect, yet still be lightweight. Notice the large amount of tubing which is required to support the rocker arm. This is a lot of weight which is placed way up front. Also notice the neat aluminum pillow block mounting for the anti-roll bar. Very nice. Note the adjustability of the bar arm length.

1

(1) Instead of bringing the rear kicker bars straight back from the main hoop of the roll cage, Chrysler cars build these bars in an "X" shape. This provides greatly improved torsional rigidity. Also note the triangulating bars which go from the center of the "X" to the frame rails at the side. This completes a fully triangulated rear bay for this car to prevent the rear spring load inputs from flexing the frame rails.

(2) This type of Panhard bar is the perfect answer for clearing the hump of a quick change. The radiused section of the tube is bent from thick wall tubing.

2

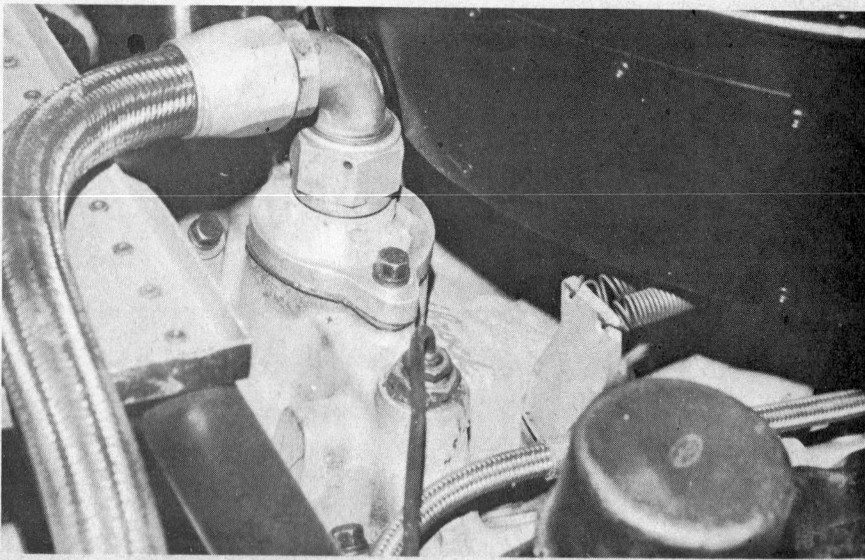

Always use two throttle return springs. And even more important, be sure the spring anchor is a very solid one (see right and bottom on page 15).

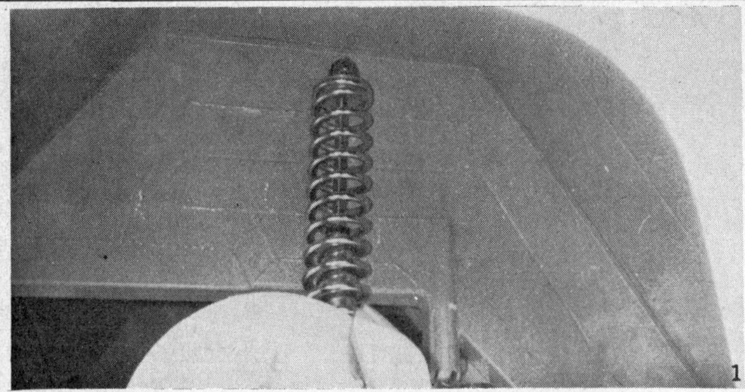

1, 2 The mount of a coil-over-shock suspension unit on the rear end must be as low as possible. This mount is too high, and the result was that the rear end was loose and no amount of spring changes could remedy it. The spring was mounted to high to resist body roll, and received it in a lateral motion instead. As a side note, look at that beautiful inner fender panel sheetmetal work.

(3) When you use the stock Chevelle trailing arms mounted straight forward, as shown here, you give your car another option for a rear spring. The mounts are in the correct position so that a Chrysler racing leaf spring can be used!

These five photos illustrate a very well-triangulated chassis. This car turned out to be very rigid, but nearly 350 pounds under minimum weight requirement. This resulted from a proper planning and placement of roll cage tube components.

A perfect way to build a race car is to suspend it off of stands such as these. It allows the versatility of placing the vehicle in any position. You can even turn the car upside down to weld or fit components (see below). If you anticipate having to move the car around, put casters on the stand (see below).

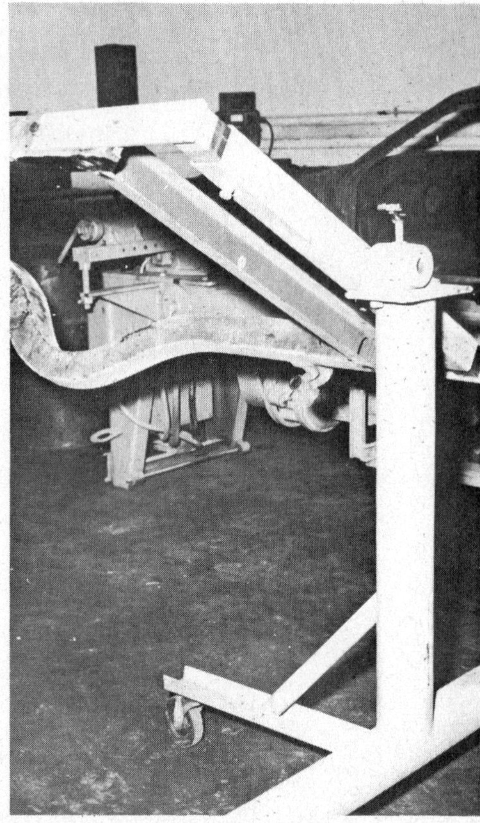

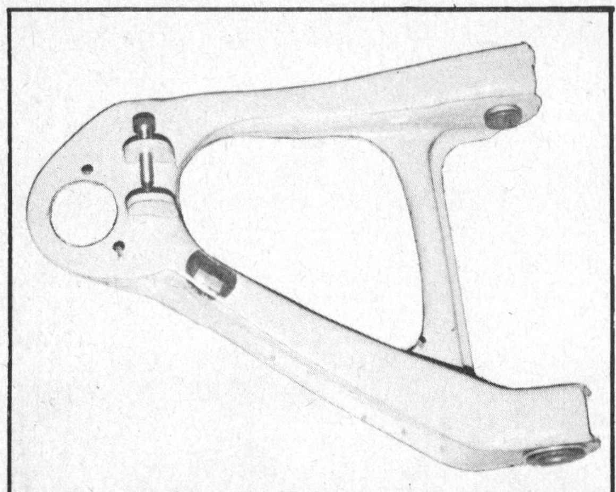

This is how the lower A-frame can be trimmed when a coil-over-shock suspension unit is used.

Vibration can chafe a hose or wire passing through sheetmetal. The way to beat that is to protect the hose or wire with a rubber grommet in the hole. Protect the driver, though, by making the hole in the firewall only as large a diameter as is necessary to pass the wire or hose. A large opening in the firewall can act like a blow torch nozzle when there is a fire under the hood.

The roll cage bar which NASCAR requires to run from the main hoop to the right front corner is a very long span, and thus it is very susceptible to buckling. Breaking up this span in a series of triangulated short spans makes it much, much stronger. It is much less liable to be bent in a minor hit at the right front. Note that the triangulating bars need only to be 1 ¼" OD material to be strong enough, thus saving weight.

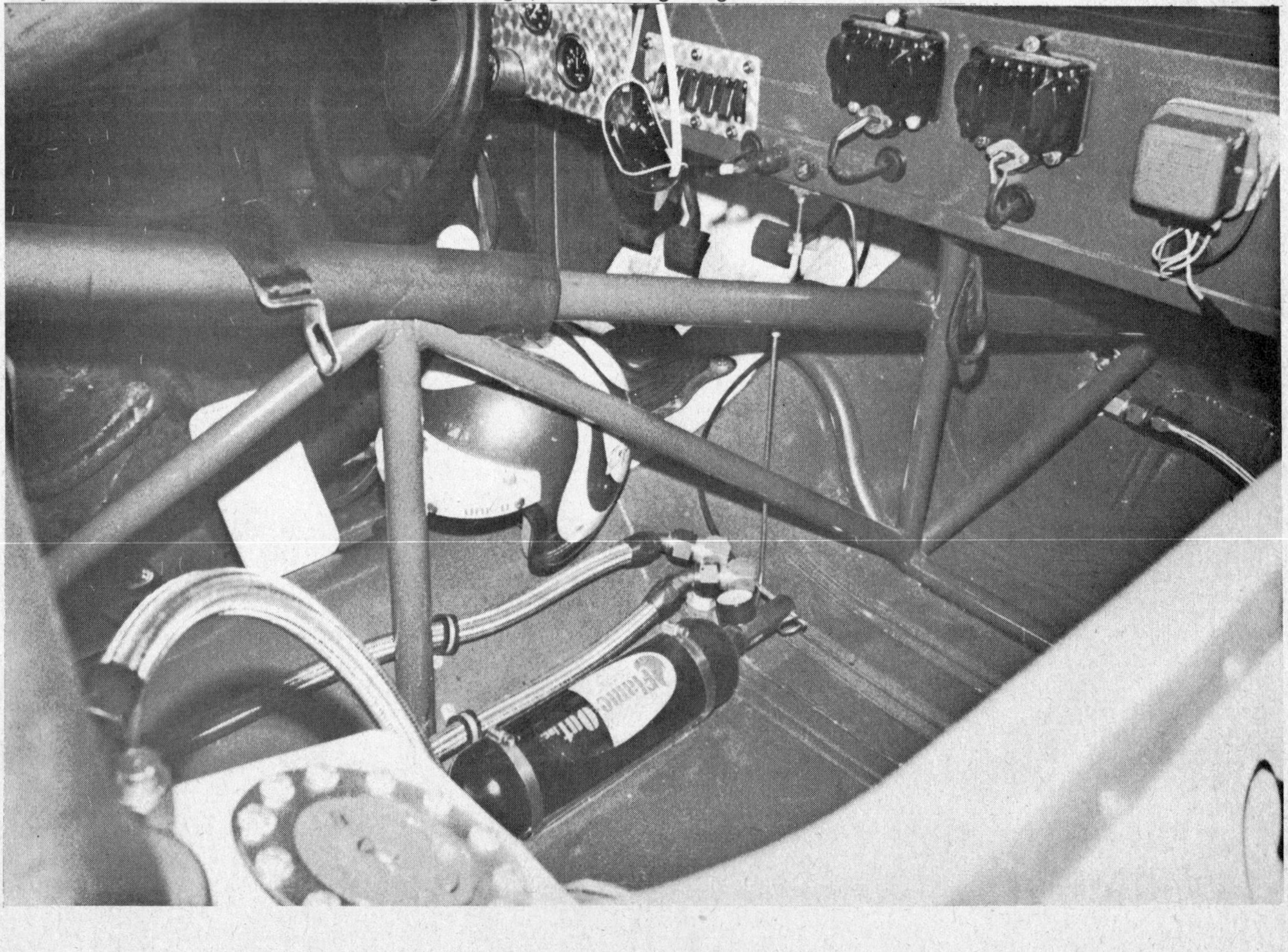

Above, if you are building a steering system or want to double check yours for parallelism, it can be done with two simple measurements as illustrated by lines A and B. Line A should be the same length as line B, measured from center of pivot to center of pivot.

(1, 2 & 3) When you're changing engines it's a pain to unbolt and re-attach the alternator wiring all the time. So, this terminal block mounted 18 inches away from the alternator is a great help. The hot lead is simply interrupted and bolted around a brass post. The rest of the harness goes through a quick-connect harness under the terminal block. The terminal itself is protected with a piece of rubber hose which slides over it. A side benefit of this terminal is that the terminal can be used for the hot lead for the timing light. Most generally on race cars, the hot lead on the battery and starter solenoid are out of reach of the timing light.

1

2

3

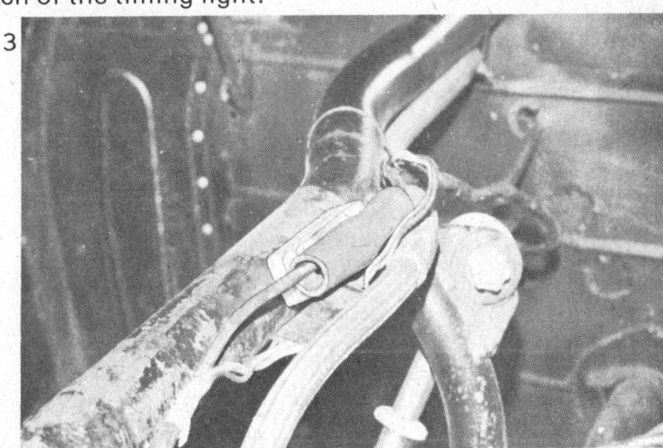

How do you cut a spring? Crank up the old cuttin' torch and let it go. Let the spring naturally air cool.

Instead of building a fixed height rear spoiler on our car, consider adding a movable spoiler which can be adjusted for height according to the track being run. This is especially helpful on a sportsman car which runs weekly short tracks (where all the spoiler you can get helps) and then one or two big sportsman races a year at places like Ontario, Charlotte or Atlanta.

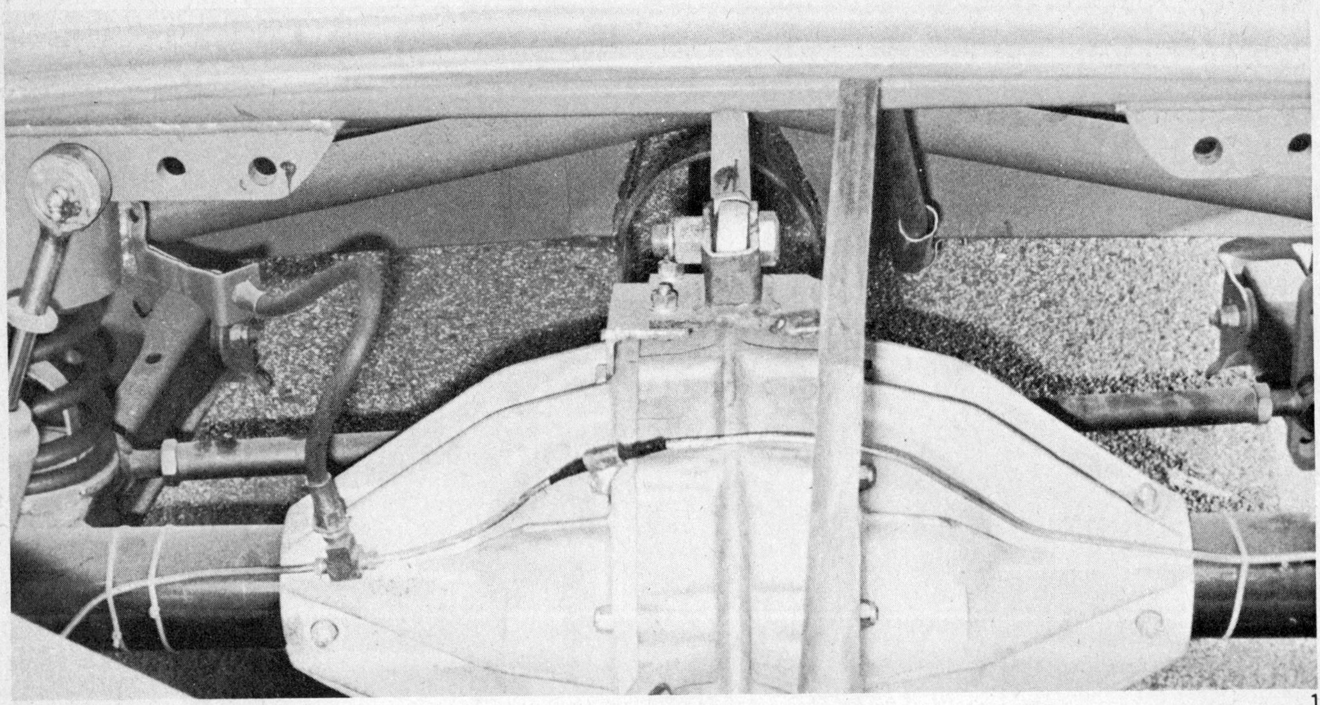
1

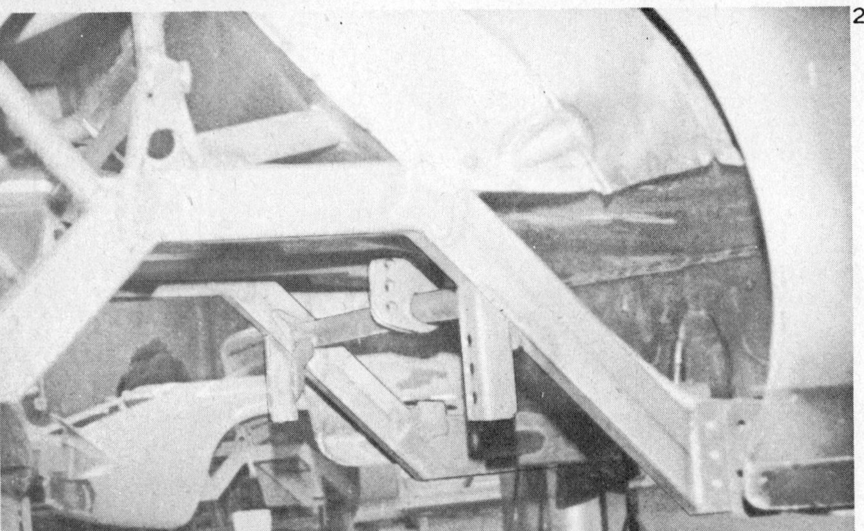
2

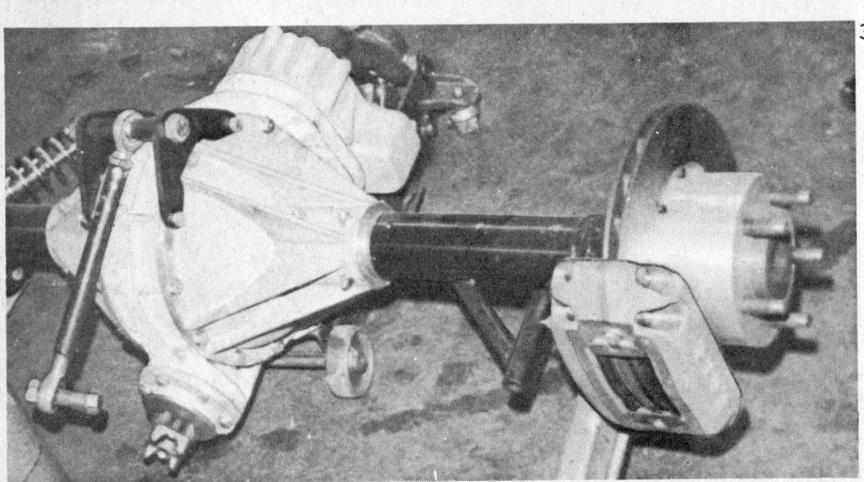
3

(1) One good way to skirt around the troublesome hump on a quick change with the Panhard bar is to run it in front of the quick change. Note the three alternative mounting holes for the shock absorber. This helps to fine tune the shock absorber rate. If less rate is required, the shock is angled inward more.

(2) If you use a Panhard bar in front of the quick change, this is the type of frame bracket to use. There are two brackets here so the bar mounting can be reversed for road courses.

(3) This is the type of rear end mounting bracket to use for a front-mount Panhard bar. Notice how the triangulation braces the mount against the direction of force.

A good electrical panel. Positive lead coming in to terminal bar is large gauge wire, all wires are routed through rubber grommeted retainers. Notice large red light on dash which lets drivers know if engine has died (for instance if he spins).

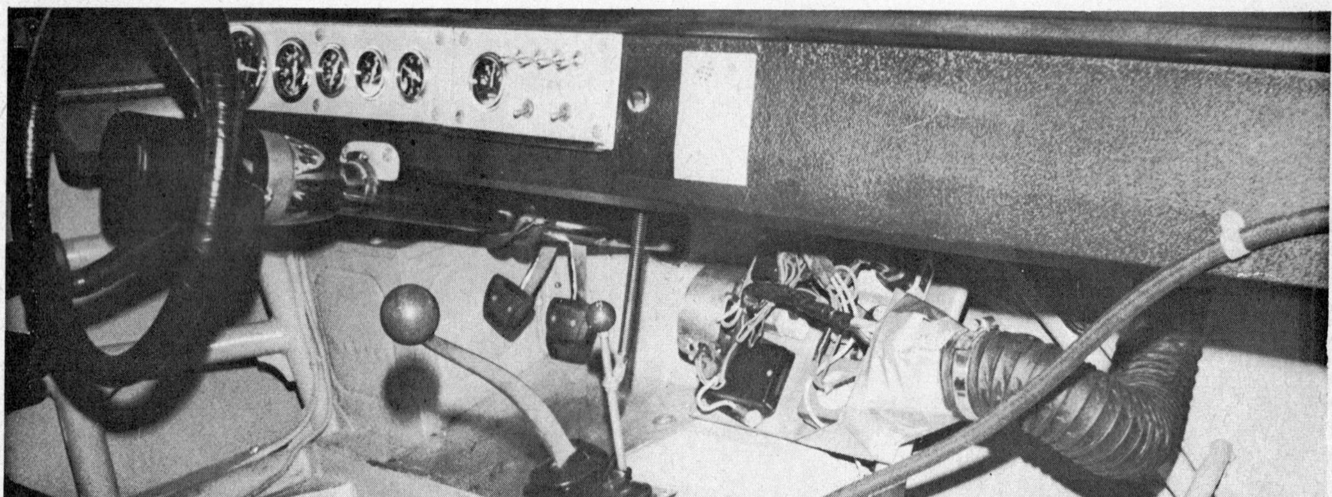

Above, no chances of ignition failure are taken here. See the hose feeding cool fresh air to the ignition parts? In all practicality, this is only needed for a long race in hot weather.

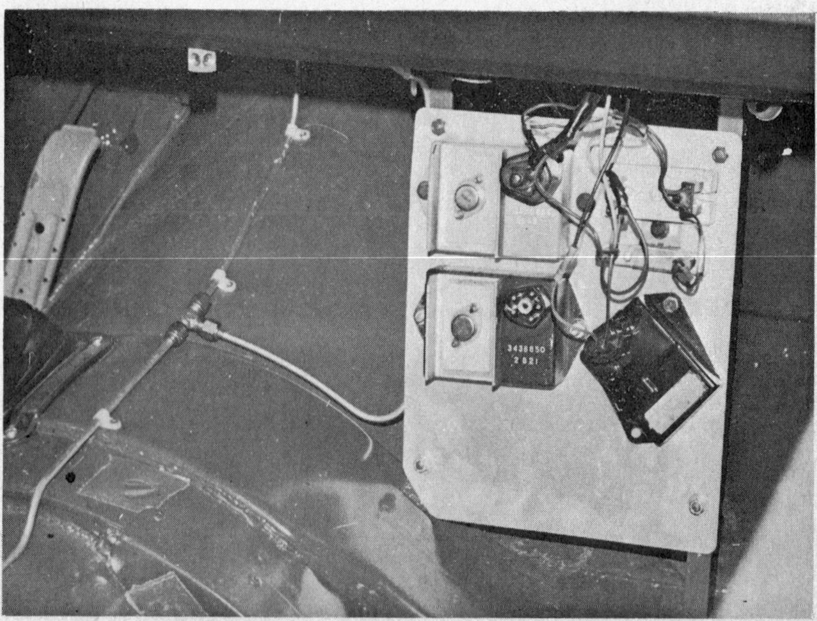

Another good ignition panel. Note that there are two electronic control boxes. If one goes out, just unplug it and plug into the other.

Notice the rubber pads the radiator sets on here. A good idea to keep vibrations out of the radiator. The ignition component was placed in the air stream here to keep it cool, but in case of any minor dents in the front end, it is very vulnerable. Better that it is moved further back, like on the firewall or inside the cockpit.

Arrow points to a very good solid engine mount. If you are going to use solid mounts, attach them at the front and/or rear of the block, not in the stock position. This prevents lateral flexing of the block material.

What a way to build the under dash world! Notice the substantial bracing of the brake pedal. Also note the non-slip surface of the clutch pedal. There is a pivot on either side of the brake pedal because it is actuating two master cylinders. Note that the brake is on the left side of the steering column. This is an important safety measure. If the driver were braking with his left foot on the pedal on the right side of the steering column, and the car suffered an impact, most probably the driver would suffer a broken leg.

Good mount of the coil-over-shock suspension unit. It is low on the axle, and close to the wheel. Note the rear anti-roll bar installation.

The bar which runs across the top of the engine is intended to keep the frame rails and the cage structure from twisting inward. It can't be too effective, however, when the bar has a radius in it. The bar will just deflect at its bend. Note the "X" bracing in the upper A-arms. It really isn't necessary, and only serves to add to the unsprung weight. Also note the hose heading rearward from the radiator overflow tank. It ends behind the rear wheels.

When the front spoiler picks up air, what happens to it? It is force-fed into the radiator with this type of ducting. Only the top half of the grille opening is used for supplemental air. This is a good trick for a wet dirt track where the grille gets packed with mud.

Above, a great way to keep the inside of the race car clean, hose it down, then open up the drain plug. Notice that the floorpan sheetmetal has been formed to place the drain at a low spot for easy drainage. Also notice the rubber grommet retainer keeping the braided hose in place.

Below, this cage structure would be stronger (to resist torsional deflection) if the other half of the "X" in the main hoop would be installed. A better place for the oil dry sump tank would be behind the driver's seat. This would help to gain more rearward and left side weight bias.

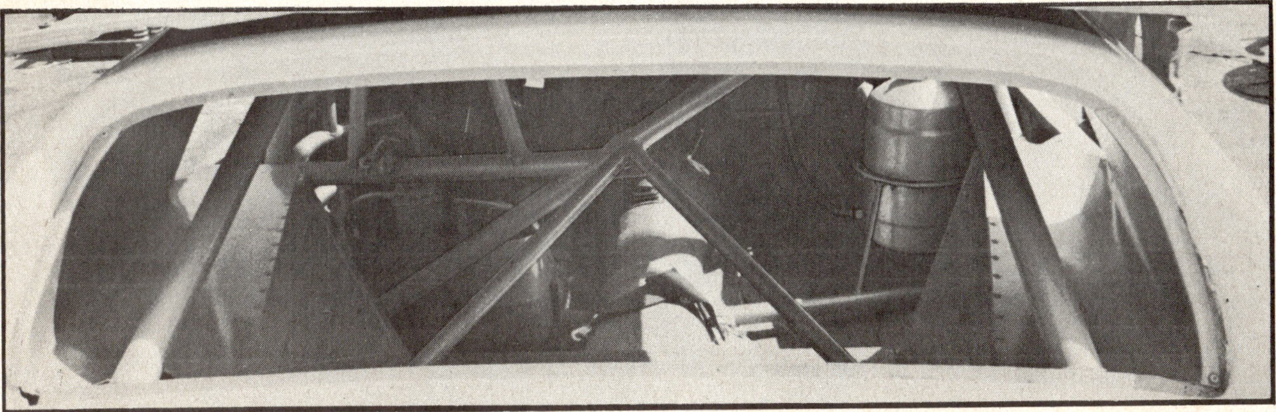

Stock floorpans in stock cars should be removed, or moved higher for greater ground clearance and driveline clearance. Good substitute floorpans can be found in a '67 Galaxy and '68 Camaro.

The new floorpan is clamped and propped in place until it can be tacked in place.

Spacer material is added between door sill and floorpan.

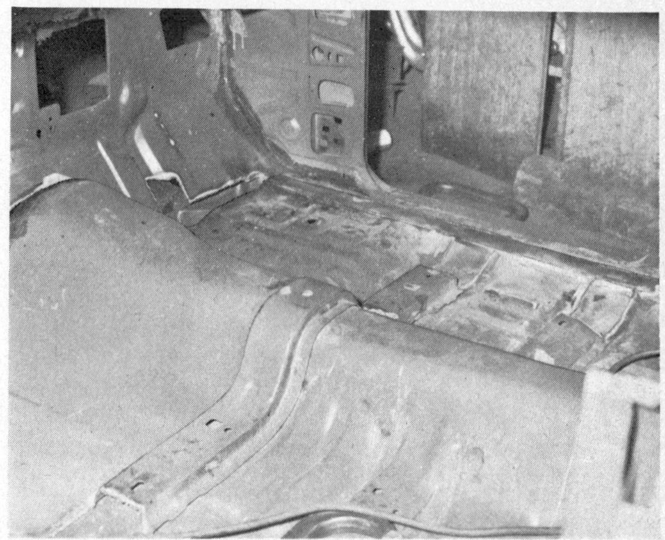

A completed installation.

Of course, if the sanctioning association rules don't specify that a stock-type floorpan be used, the easiest floorpan is made from a flat piece of sheetmetal.

A '67 Ford Galaxy floorpan mounted in a Chevelle. Notice that the floorpan is welded to one side of the frame rail, and the doorsill is welded to the other side of it.

This frame crossmember is not nearly strong enough to support the input loads of the front suspension. There should be a piece of rectangular tubing fully connecting the two halves of the crossmember.

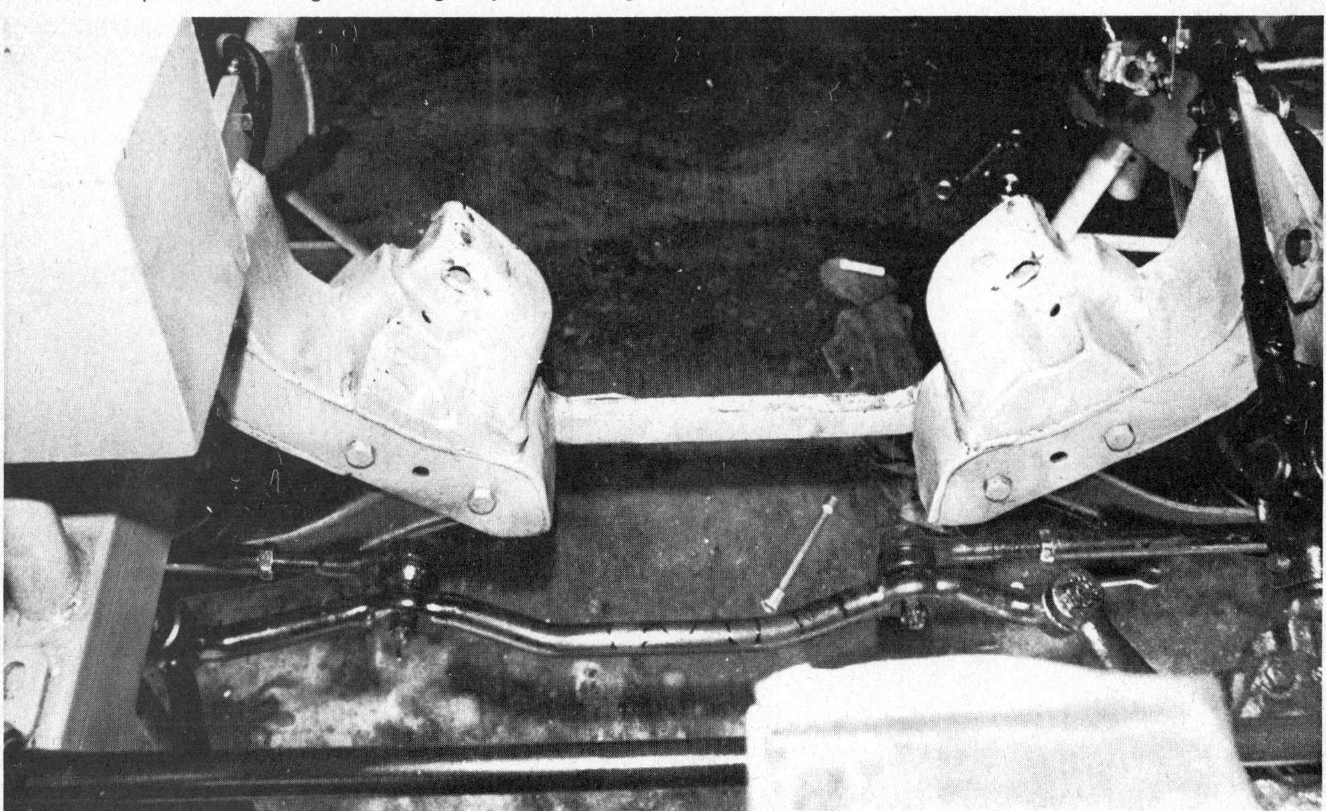

This battery is in the correct position, but it is not protected by a box or shield of any kind, and is very vulnerable.

Notice where the battery is? It's down low, all right, and far back, but not on the left side. For oval track racing, the battery is one item which can be located on the left to gain the preferred left side weight bias. Perhaps the driver is afraid of the vulnerability of the battery on the left, but a protective shell can be built around it.

Good location and good shield for the battery.

A radiator shroud definitely is a big advantage in getting the most in cooling power from a radiator. Make sure it has a tight mount against the radiator.

This radiator is so far away from the fan that the fan will have virtually no effect on it. A fan spacer and a shroud might help if there is an overheating problem. Otherwise, a larger size radiator would be needed (because it doesn't appear that the radiator can be moved much further back). Notice the neat vibration-free mount of the radiator.

Attention to detail always pays off. See the cotter pin which was not secured on the brake caliper and worked its way out?

A sensibly-designed front end structure. Triangulation supports the hood pin and the forward end of the fender. Triangulated structure is welded to a piece of ½" x ½" square tubing which lines the inside of the fender. Piece of square tubing from frame rail supports bumpers. It is placed on an angle so a bumper-bending crash won't necessarily bend the frame rails.

Below, the reservoir has been extended on the master cylinder to provide extra brake fluid for disc brake operation. The fluid line leading out of the master cylinder is steel braided. It must be that or steel tube in a race car application. Note the hood spring—these are available from companies which make catering truck bodies.

If you are going to make your own interior sheetmetal, it is best to make a template first out of cardboard.

A clean, uncluttered engine compartment. The steel braided hose is strong, looks great, and is quite reasonable when purchased at a surplus store. All the lines are held in place with plasti-ties. Note the piece of "racer's tape" between the bottom of the windshield and the firewall. It helps the flow of outside air into the engine compartment (which feeds the carburetor).

Below, the radiator top attachment plus fan shroud attachment are designed very well. Radiator is from a Corvette. Notice the shielding built on either side of the engine breathers. This prevents the fan from blowing dirt into the breathers, or oil from blowing into the air filter.

The rib around the upper perimeter of the upper A-arm very effectively reinforces its rigidity.

The Jabsco "water puppy" is by far the best lube pump for rear end and transmission cooler. They are available through marine equipment dealers. Be sure you buy the model with the metal impeller.

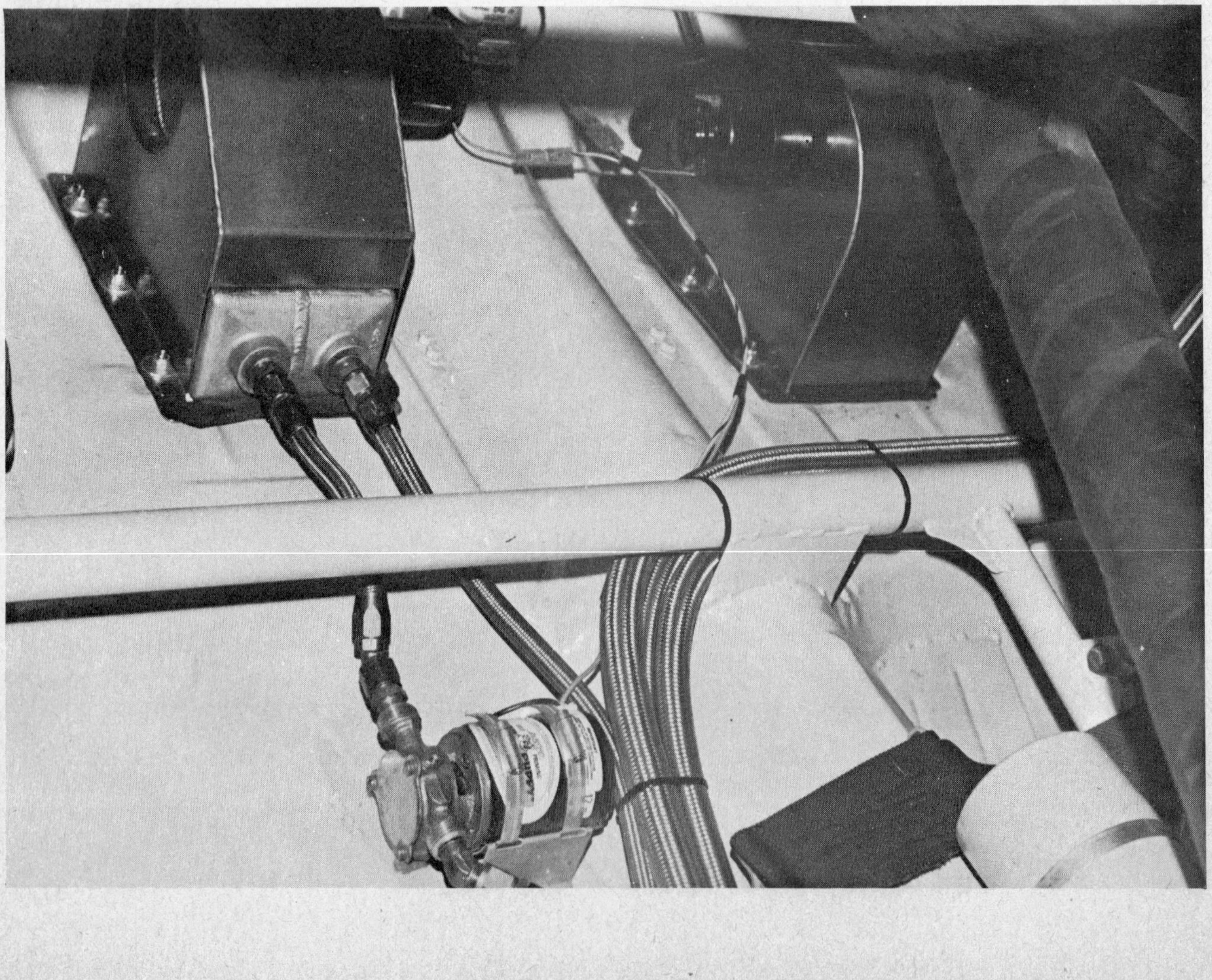

Above, the fan was not close enough to the radiator, so a spacer was used.

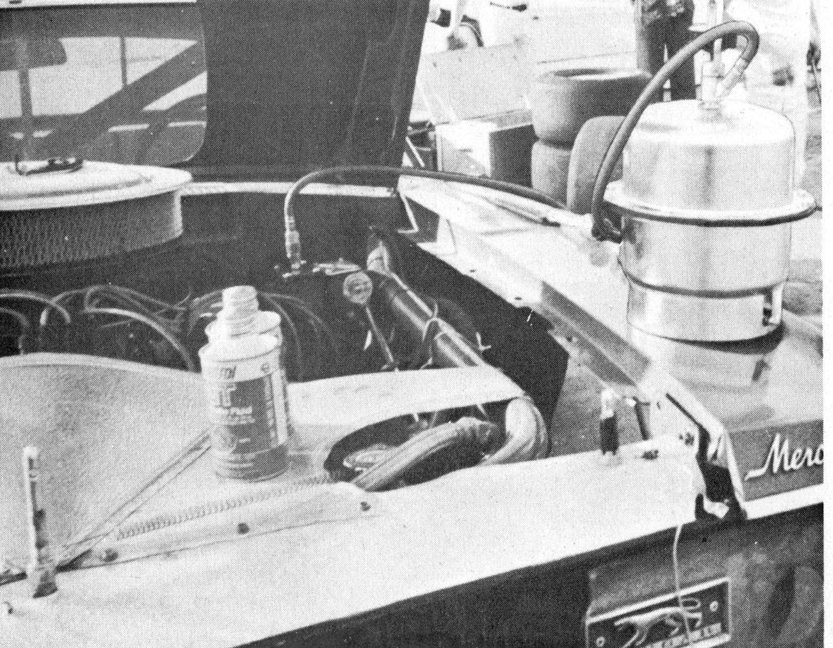

The best way to bleed disc brakes is with a pressure bleeder. This one uses a special master cylinder cover in which a fitting is adapted. Be sure pressure bleeder always has clean, new fluid in it.

The multiple attachment holes for the upper trailing arm can help tailor anti-squat. This suspension employs a trailer bar spring, but it is obscured from view by the lower trailing arm.

Look at the beautiful "X" member which extends over the engine bay and behind the firewall. This really lends structural rigidity to the chassis.

Note the bracing tube which runs between the upper A-frame mounting brackets. This adds substantial rigidity to the front suspension. Unfortunately, it just isn't practical to use on all cars.

Above, if roll cage triangulation bars are in the way of engine installation in the engine bay, make them bolt-in components such as these.

The rear end lube radiator and pump are mounted on the left side to gain more left side weight bias. Note the rubber grommets protecting the lube hoses passing through the rear bulkhead.

Above, notice how the oil filter remote adapter is mounted to the cage tube? The bracket is welded to the tube, rather than the tube drilled for attaching bolts. When a hole is drilled in a cage tube, it places a stress point in that tube. We've seen several instances of holes in tubes which have caused a split in the tube emanating from the hole. Notice the brace off of the alternator bracket which holds the top radiator hose in place.

A very well triangulated fender and sheetmetal corner brace.

An alternative idea to the Panhard bar with a quick change is this one which runs at a 60-degree angle to the rear end. The advantage to this method is that the bar can be mounted under the driveshaft yoke, giving a lower rear roll center.

Let's say you've already got the rear weight bias you want, but need more left side bias, and you've got to add more weight ballast. Where do you put it? Right in front of the driver's seat, where it can be bolted to tabs welded onto the seat frame. The tubing on the floor (see arrow) contains the fuel line. It is a ¾" ID tube.

Below, very good engineering in mounting of the steering wheel and shaft. The beauty of the shaft design is that it will not spear straight back into the driver in case of severe front end collision. Rather, it will fold up around the U-joint.

Individual wheel scales (more commonly called grain scales) like these are the best way to determine the wheel weights at each corner. They are expensive, but four racers might consider buying them in partnership (one apiece). The other alternative to the wheel scales is the wheel load checker or chassis checker. An advantage of using four wheel scales is that it allows the racer to move weight around in his car and see what effect it has at each wheel.

A nice radiator installation. Note the mounting tabs on top which are an integral part of this custom-made radiator. The tabs are a frame soldered between the core and tanks. There is rubber insulation material between the tabs and the mounting tube to protect the radiator from vibration.

The first class way to brace your windshield. These braces are made from aluminum stock. They have foam rubber insulation between the glass and the aluminum.

Above, the quick-connect fittings on all wires allow very fast engine changes. Notice the eccentrics in straddle-mounts of the upper A-arm to allow front end alignment changes. This is a good method, but it requires several different upper A-arms to have a range of adjustability. The eccentrics do not afford a lot of adjustability.

The master cylinder/brake balance bar unit represents a very unwieldy piece hanging off the firewall. In fact, its flexibility can cause a soft pedal. To remedy this, a bracket should be fabricated which attaches to the front of the balance bar bracket and ties over to the tube running next to it.

This nerf bar is braced to the main front cage upright. If it takes a moderately hard knock, these uprights, the main front hoop, and the radiator ducting will be bent. Also, the front tube uprights support the braces which hold the hood pins, and front sheetmetal, all of which would probably be bent as well.

This nerf bar is mounted to brackets off the frame rails. If it suffers a hard blow, none of the front cage tubing or sheetmetal supports will be affected. Most probably the bar's attachment brackets and the bar itself will bend and yield before the frame rails ever get bent.

Notice the shock absorber mounts? They are not substantial enough to handle either severe load inputs or minor vibrations. In either case, they will deflect. A shock absorber should be mounted to a very rigid, sturdy structure, such as the top cage tube in the engine bay.

These coil-over-shock suspension units are well braced in the chassis, but only in one direction (inward toward each other). The roll cage tube which ends on the frame just behind the suspension units should have been brought to the tube structure supporting the suspension units. This would aid in supporting input loads in the fore and aft direction.

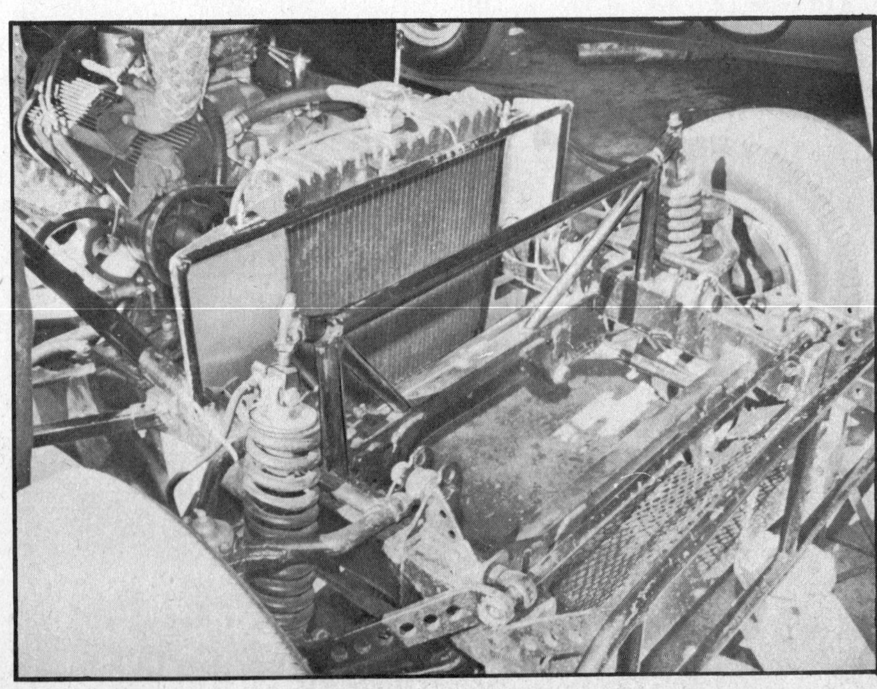

This is probably the best type of upper A-arm to use with straddle-mount brackets. It has a wide range of adjustability, is easily adjusted, and is strong.

The two pieces of 1¾" OD tubing which brace the 90-degree joints in the frame rails add greatly to the torsional strength of the frame. The pieces are simple to install, don't weigh much, but add a whole lot to the overall product.

Above, an interesting trailing arm system—parallel trailing arms running in opposite directions. This system should work out o.k. if the inner ends for the arms are the same distance above and below the axle, if the arms are the same length and run parallel to each other and the ground at running height.

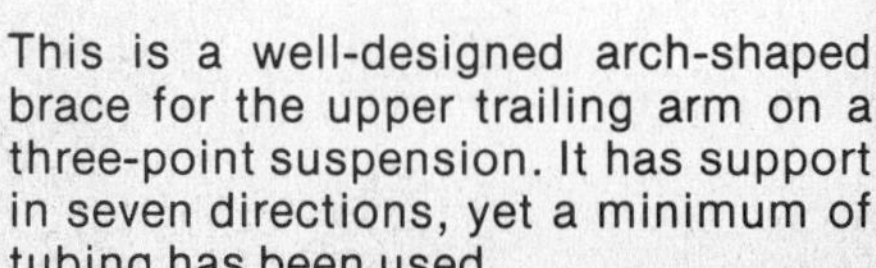

This is a well-designed arch-shaped brace for the upper trailing arm on a three-point suspension. It has support in seven directions, yet a minimum of tubing has been used.

This group of photos depict two cars, a Nova and a Camaro, built by Ivan Baldwin. The chassis structure was designed by Baldwin, and is being manufactured and sold in kit form by McCoy Racing Products in Modesto, California. The chassis kit is particularly designed to be used with Carrera coil-over-shock suspension units.

This photo shows the engine bay of Baldwin's Nova. Note the cage triangulation where the suspension unit attaches. Also note how the engine breathers are protected from the fan turbulence.

The base for the front subframe of the Baldwin car is an early Camaro (or '70-'73 Nova) subframe. The major tubular component of the cage structure in the engine bay is bent in an arch shape. This is a very efficient and strong shape. To support the suspension input loads in all directions, the arch is braced fore and aft with two triangulated tubes and laterally with one removable tube.

The front frame crossmember is trimmed just inside the lower A-frame forward mount, and the anti-roll bar support tube replaces it. Saves weight and space.

Baldwin builds the upper A-frame mounts to include nearly 50% anti-squat.

The complete Baldwin-built chassis.

Above, notice the cage tube which protects the battery, the "X" member in the front (which adds greatly to torsional strength) and the lightweight brace plates on the upper A-arm attachment bracket.

See the tube which runs from the bottom rear of the right side door to the "X" under the dash? Baldwin found this bar added more to torsional rigidity than any other. Another bar Baldwin found to be a great aid to strength was the diagonal above the driver's head.

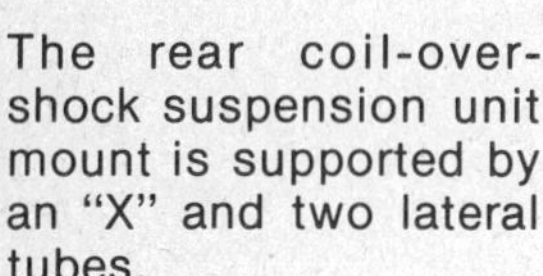

The rear coil-over-shock suspension unit mount is supported by an "X" and two lateral tubes.

This is a very good strong engine bay cage structure on a modified to support the input loads of the coil-over-shock suspension units.

LIST OF SUPPLIERS WE MENTION

McCoy Racing Products
Jack McCoy and Ivan Baldwin
553 S. 7th St.
Modesto, CA 95351
[209] 521-5656

Petty Enterprises
Rt. #3
P.O. Box 621
Randleman, NC 27317
[919] 498-3745

Speedway Engineering
Frank Deiny
2471 Fletcher Drive
Los Angeles, CA 90039
[213] 666-9510

Tex Enterprises
Tex Powell
Rt. 4, Box 348
Asheboro, NC 27203
[919] 625-3943

Other books from Steve Smith Autosports

The Stock Car Racing Chassis

A basic handling and suspension book for the oval track racer. It contains practical and basic information on the design and adjustment of a racing chassis. It also covers such topics as caster, camber, toe-out, reading tire temperatures and basic chassis problem sorting. A special feature is the well-illustrated chapter on building a competitive Chevelle chassis.

No. S101 $5.50

The Complete Stock Car Chassis Guide

This is an advanced and revealing handling and suspension book which was researched by way of consultations with the sport's foremost chassis experts. Contents include racing tire principles and selection guide, dirt track chassis design, roll cage construction, shock absorber selection, how to build the two-point rear suspension, aerodynamics, building the MoPar chassis, roll couple distribution, anti-roll bars, and much, much more. Packed with photos and drawings.

No. S102 $5.50

The Trans Am and Corvette Chassis

Detailed handling, suspension and chassis design information for the Corvette, Mustang, Camaro and Firebird chassis. Text includes bump steer, chassis building, roll couple distribution, racing tire principles and selection, Koni shock absorber selection and use, reading tire temperatures, and much more. Written with the help of famed road racer and engineer Dick Guldstrand.

No. S103 $5.50

Stock Car Driving Techniques

Every aspect of competition driving is covered in this exciting 96-page book written by 1973 NASCAR Grand National champion Benny Parsons. *Southern Motoracing* reviewed it saying, ". . . it is rich in the kind of information that can help the racing fan to enjoy and appreciate what he sees on the speedways. It is a wiser investment, indeed, for the thousands of weekend stock car racers all over the nation who could learn a great deal from the knowledge imparted by Parsons."

No. S104 $7.45

Advanced Race Car Suspension Development

This book offers the latest technological information about race car chassis design and development. It supplies you with all the knowledge you need in order to precisely predict what your car will do under any specified circumstances. A former GM chassis design engineer said of this book: "This is the most complete and accurate suspension book that has ever been published." Another famed racing team engineer said: "This book offers more information than all the combined books I know of. This is definitely a book everyone in racing can use." Contains 176 pages, 100 photos and drawings.

S105 $6.45

Work Book for Advanced Race Car Suspension

This Work Book was designed to accompany our "Advanced Race Car Suspension" book. It extracts all of the formulas presented in the book and presents them in an organized, point-by-point manner so you may follow them as examples for your car. It also contains two examples of a complete chassis analysis on sample cars.

No. WB5 $3.45

Racing Engine Preparation

From building to tuning, this book covers every aspect of race car competition engines. Waddell Wilson, the author, has been the engine building genius behind such stars as the Holman and Moody team cars, Bobby Allison, Benny Parsons, and A.J. Foyt. Says Foyt, "This is the most thorough engine book you'll ever read. Wilson has put his 20 years of engine building experience into it, and he is a man I thoroughly trust for my engines. You're missing something if you don't have this book." The goal of this book is to help you get the most in power and reliability from your engine, and for the least dollars. Some of the fully illustrated chapters include: A Complete Guide to Camshafts • The Lubrication System • The Ignition System • A Complete List of Major Part Numbers • Detailed Engine Assembly • Dyno Testing • Plus, complete chapters on the preparation of pistons, rods, carburetors, blocks and much more.

No. S106 $8.00

Race Car Chassis Blueprints

Chevelle Chassis blueprint shows all views and dimensions for building a competitive GM chassis. Includes part numbers, materials, and construction notes.

No. B301 $10.00

Square Tube Chassis blueprint details the latest in stock car technology, the chassis completely fabricated from square tubing. It details all views and dimensions, gives part numbers, materials and construction notes. Everything you need to know.

No. B350 $10.00

Camaro Full Frame chassis is based on using the 1970-74 Camaro stub frame attached to a full length square tubing frame. Print shows all views, dimensions and parts required for a top quality chassis.

No. B375 $10.00

Tech Tips

A monthly question and answer technical newsletter designed to be your one-stop informative source for all race car problems, be it chassis, handling or engines. Our panel of experts are ready to handle any inquiry, plus pass along any new developments.

No. TT-1 $10 a year

To order any book listed, specify book number and send full price to:
Steve Smith Autosports, P.O. Box 11631, Santa Ana, CA 92711

Race Car Braking System

All about disc and drum braking systems. Which brake fluids are best? What's the difference between silicone and glycol brake fluids? Which passenger car brake drums can you use for racing cars? What are their part numbers? Which passenger car disc systems can you use for racing? Where can you find lining for these? Which size master cylinders should you use? How can you increase the braking system line pressure? What size wheel cylinders should you use? These questions and more are answered in-depth in this book.

No. S107 $5.45

The Racer's Complete Reference Guide

This "bible of the high performance industry" was put together so a racer could get it together with less hassle. He can spend more time racing and working on his car and less time trying to find out where to get this and how to find out that. Covers everything ever needed for a race car. Where to get: hardware, chassis and engine parts, running gear, racing rules. Where to find: fabricators, engine builders, safety gear, racing associations. Listings of: companies, addresses, phone numbers. This brief description just scratches the surface of this book's contents.

No. S108 $5.50

High Performance with Fuel Injectors

What do you want to know about buying, installing and tuning fuel injection systems? This book gives it all to you, straight and factual. Covers fuel pumps, metering valves, nozzles, by-pass valves, cut-off valves, fuel filters, fuel pump drives, ram tubes, fuel lines and tanks, how to install injectors, initial starting procedure, maintenance, plus a complete trouble shooting chart. Very well illustrated.

No. S109 $5.95

If You Desire The Ultimate In Performance,
Our Technical Books Are

THE SOURCE!

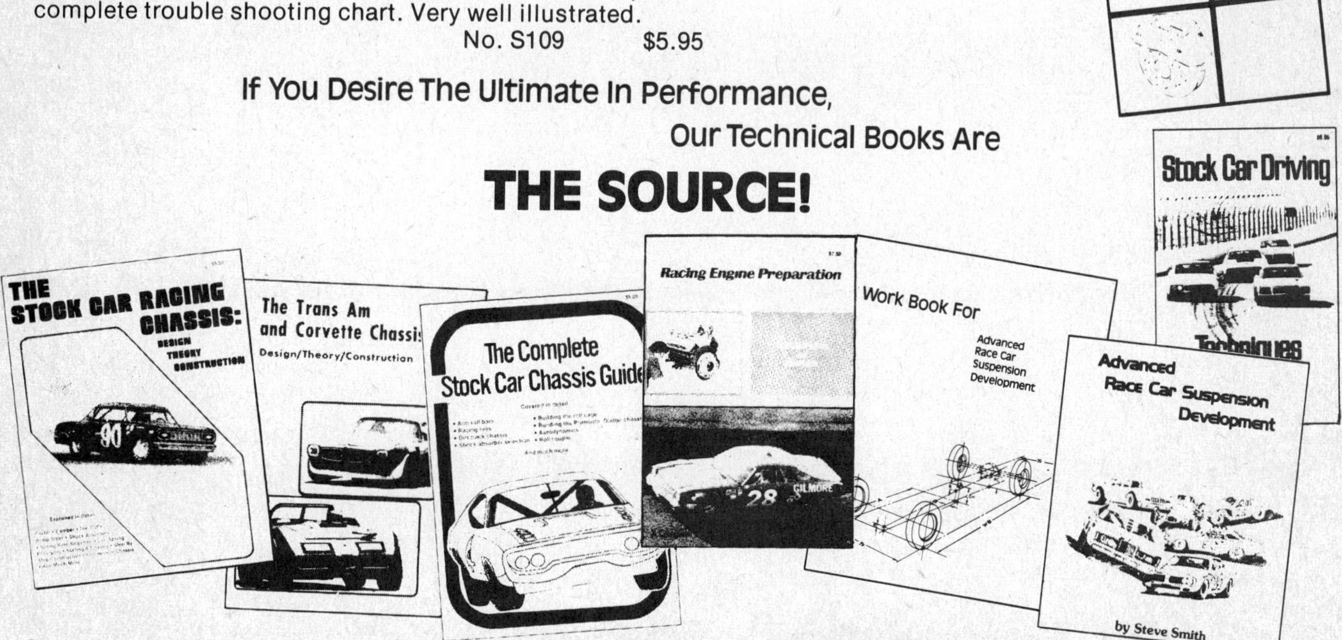